Environmental Awareness:
AIR POLLUTION

AUTHOR
Mary Ellen Snodgrass

EDITED BY
Jody James, Editorial Consultant
Janet Wolanin, Environmental Consultant

DESIGNED AND ILLUSTRATED BY
Vista III Design, Inc.

BANCROFT-SAGE PUBLISHING, INC.
601 Elkcam Circle, Suite C-7, P.O. Box 355
Marco, Florida 33969-0355

Library of Congress Cataloging-in-Publication Data

Snodgrass, Mary Ellen.
   Environmental awareness—air pollution / by Mary Ellen Snodgrass; edited by Jody James, Editorial Consultant; Janet Wolanin, Environmental Science Consultant; illustrated by Vista III Design.
   p. cm.—(Environmental awareness)
   Includes index.
   Summary: Discusses the different types of air pollution and how they can be stopped or reduced.
   ISBN 0-944280-31-5
   1. Air—Pollution—Juvenile literature. 2. Air quality management—Juvenile literature. [1. Air—Pollution. 2. Pollution.] I. James, Jody, Wolanin, Janet. II. Vista III Design. III. Title. IV. Title: Air pollution. V. Series: Snodgrass, Mary Ellen. Environmental awareness.
TD883.13.S66   1991
363.73'92—dc20

International Standard Book Number:
Library Binding 0-944280-31-5

Library of Congress Catalog Card Number:
90-25726
CIP
AC

PHOTO CREDITS
COVER: Vista III Design; Nancy Ferguson p. 38; K.G. Melde p. 4, 29; Minnesota PCA p. 6; Holly Neaton p. 42; Northstar Photography, Bob Paulson p. 11; Unicorn Photography, Eric Brandt p. 21, Martha McBride p. 30; Vista III Design, Ginger Gilderhus p. 14, 17, 36, Grant Gilderhus p. 12, 18, 23, 25, 26, 27, 29, 33, 35, 40.

Copyright © 1991 by Bancroft-Sage Publishing, Inc. All rights reserved. No part of this book may be reproduced in any form without written permission from the publisher, except for brief passages included in a review. Printed in the United States of America.

TABLE OF CONTENTS

CHAPTER ONE — **AIR TO BREATHE** 5

   A Trip to Aunt Grace's House 5

CHAPTER TWO — **AIR AND YOU** 9

   Air And Lungs 9
   Pollution And Disease 10
   Sampling Air Pollution 13

CHAPTER THREE — **DIFFERENT TYPES OF AIR POLLUTION** 15

   Natural Pollution 15
   Direct Tobacco Smoke 16
   Secondhand Smoke 16
   Pollution From Vehicles And Factories 19
   Acid Rain 23
   Indoor Pollution 24
   Pollution From Farms 27
   Other Polluters 28

CHAPTER FOUR — **CLEANING THE AIR** 31

   Stopping Air Pollution At Home 31
   Making Our Nation A Better Neighbor 32
   Reducing Pollution From Cars 34
   Making Safer Factories 34

CHAPTER FIVE — **THE INDIVIDUAL'S PART** ... 39

   Protect The Air 39
   Protect Yourself 40
   Encourage Government Officials
      To Take Action 42
   Encourage Industry To Keep The Air Clean 42

GLOSSARY 43-48

Air pollution harms people in the country as well as in the city.

CHAPTER ONE

AIR TO BREATHE

Air is one of nature's greatest treasures. It forms a layer less than twenty miles deep that circles the earth. Animals and plants need this layer of air to live. Without air, all living things would die. In a short time, an airless earth would be as empty and silent as the moon. Our blue sky would be black.

If we want the world to be a comfortable home in the future, we must protect the air around us from harmful **pollution**. In many places, unfortunately, the job has become difficult. Too many families crowd into cities. Too many cars fill the highways. Too many smokestacks pour harmful gases into the air.

Air pollution from these sources does not stay in one place. It drifts across rivers and fields and across the borders of countries. It mixes with rain and returns to the land below. Air pollution harms people in the country as well as in the city. Here is how air pollution affected the Grahams, who were traveling to California for a family visit.

A TRIP TO AUNT GRACE'S HOUSE

"Couldn't we go a little faster, Mom?" Jenny whimpered. "I'm thirsty. Is it much farther to Aunt Grace's? We've been driving since seven this morning."

Ann Graham stopped humming, turned on her signal light, and moved to the right lane of traffic. "We're getting off the interstate right now, Jen. Be patient."

Marty, Jen's older brother, opened both eyes and looked around. "We're in California, aren't we, Mom?"

"We sure are. As soon as we get through the Pasadena traffic, we will be on our way to Los Angeles and Aunt Grace's house," Mrs. Graham replied.

"What time is it? It sure seems dark," Marty said. He pressed his face against the car window and looked up at the sky. Hazy clouds filled the air around the tall buildings. It was hard to see the sun.

"It's only two o'clock, Marty," Ann replied.

"Then why is the sky so gray?" he asked.

"That's **smog**, Marty. California is famous for its hazy skies. The shape of the mountains, the nearness of the sea, and changes in temperature join with air pollution to cause smog. People drive so many cars and use so many air conditioners that, when conditions are right, low clouds of dirt and gas hang in the sky."

Jenny giggled and held her doll up to the glass. "Smog rhymes with frog and dog and log," she chanted.

"That's right, Jen. The word isn't new, either. It has been around nearly a hundred years," her

City smog can turn a bright sunny sky to gray and hazy.

mother explained. "Smog is a combination of smoke and fog."

"Is it dangerous, Mom?" Marty asked. "Will the smog make us sick?"

"We might have a little harder time breathing, but we're only going to be here for a few days, Marty. I don't think we have to worry about serious damage. But for people who live here all the time, smog is a real problem. It makes some people cough. It causes other people to suffer **allergies**. Also, it makes lung diseases worse. Smog even eats away the paint on cars and houses," Ann added.

"Why doesn't somebody make it go away?" Jenny asked, frowning up at the sky.

"Well, Jen, that wouldn't stop the problem. You see, people make more pollution every day. Getting rid of one day's pollution is just a start. The secret to clean air is keeping it clean all the time." Ann studied the road signs, then turned toward Montevista Park.

Marty smiled with confidence. "I'm glad we live out in the country in a clean state like Colorado. We don't have to worry about smog."

"Well, Marty, that is only half true. We do live out of the city and most of Colorado is very clean, but we aren't completely safe from air pollution," Ann replied.

"What do you mean, Mom? Is smog going to start soon where we live?" Marty seemed worried.

"Well, no, maybe not smog. But there are many kinds of air pollution. Some kinds, such as smog, are visible. Some kinds are invisible. Pollution of all kinds floats in the air. The wind can carry it many miles. The fumes from cars and factory smokestacks in one place can affect people far away. Nobody is completely free from harmful air pollution," Ann concluded.

"Will it hurt Aunt Grace?" Jen asked.

"Nobody knows what years and years of air pollution will do to our lungs," Ann replied. "But for the sake of the earth and for Aunt Grace, we need to do something about it fast."

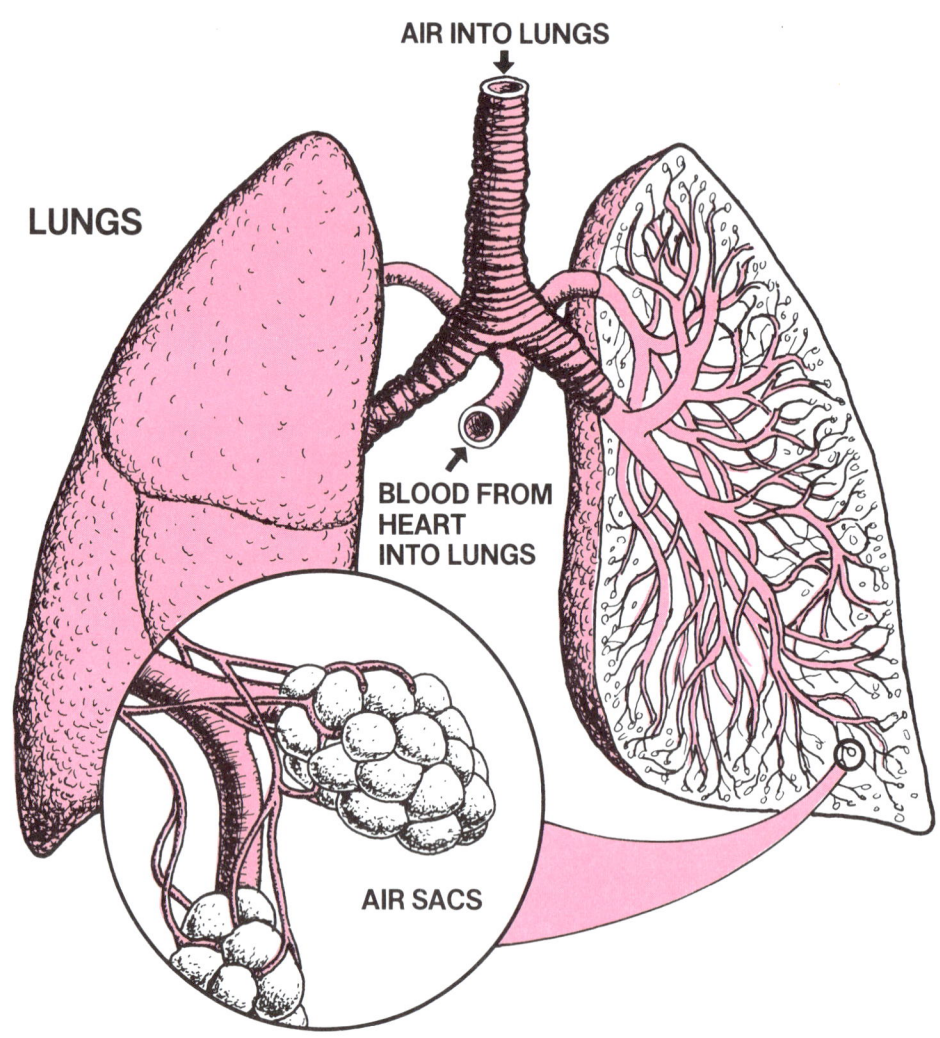

When irritating particles in the air enter the lungs and cause breathing problems, it also creates a strain on the heart.

8

CHAPTER TWO

AIR AND YOU

Wherever you live, you must breathe the air. You need to think about whether the air is clean and if not what it may be doing to your body. Is air pollution a serious problem in your area? Do you see dirty clouds forming in the sky? Is there a haze over busy highways? Do smokestacks send out streams of particles and strong fumes? How can you protect yourself from other gases that you can't even see? These and other questions will help you focus your attention on the problem of air pollution.

AIR AND LUNGS

Humans are designed to draw air into their bodies. The air, filled with **oxygen**, enters tiny sacs in the lungs. Then the air mixes with blood, which carries the oxygen to all parts of the body. Most important, blood feeds oxygen to the brain, which is the control center of the body. On its way back to the lungs, the blood carries away waste gas or **carbon dioxide**. This two-way system helps the body function and stay clean.

Several types of inner protectors help keep the body free of disease. For example, sticky **mucous linings** in the nose and throat trap dirt and smoke from the air the body breathes. Tiny hairs in the nose catch some harmful particles. Also, coughs, sneezes, tears, and runny noses wash away irritating particles.

Inside the lungs, large cells digest invaders or force them out. Together, these protectors remove some of the air **pollutants** that people breathe.

POLLUTION AND DISEASE

Irritating particles in the air cause problems for the skin and eyes, but they do even greater harm to the lungs. Strong chemicals in air pollution can burn the linings of air sacs in the lungs. They can also cause scarring or swelling, which decreases the flow of necessary oxygen in the lungs. Some particles scratch and tickle the lining of the nose and throat, which causes sneezing and coughing. As a result, the body produces more **mucus** to line the nose and throat.

These changes in the body cause people to feel tired. The added coating of protective mucus makes extra work for the lungs. These coatings narrow the passages to the lungs making breathing more difficult. All the coughing and sneezing weakens the muscles and wastes energy. In some places where air pollution is very heavy, people cannot work a whole day outdoors. Some people, especially children and athletes, should not play or run outdoors. In places where smog is unusually heavy, elderly people, babies, and people with weak lungs may even need to wear masks over their noses and mouths. These masks trap particles from smog before the pollution can reach the lungs and do damage.

When the lungs work harder to bring oxygen to the blood, the heart also must pump harder. Some kinds of air pollution cause the blood vessels to become tight. Then it is even more difficult for the heart to pump blood to and from the heart. Any problem that affects the lungs, therefore, overworks the heart. In the United States and other countries, heart disease is the number one killer. One way to protect the heart from strain is to make breathing easier.

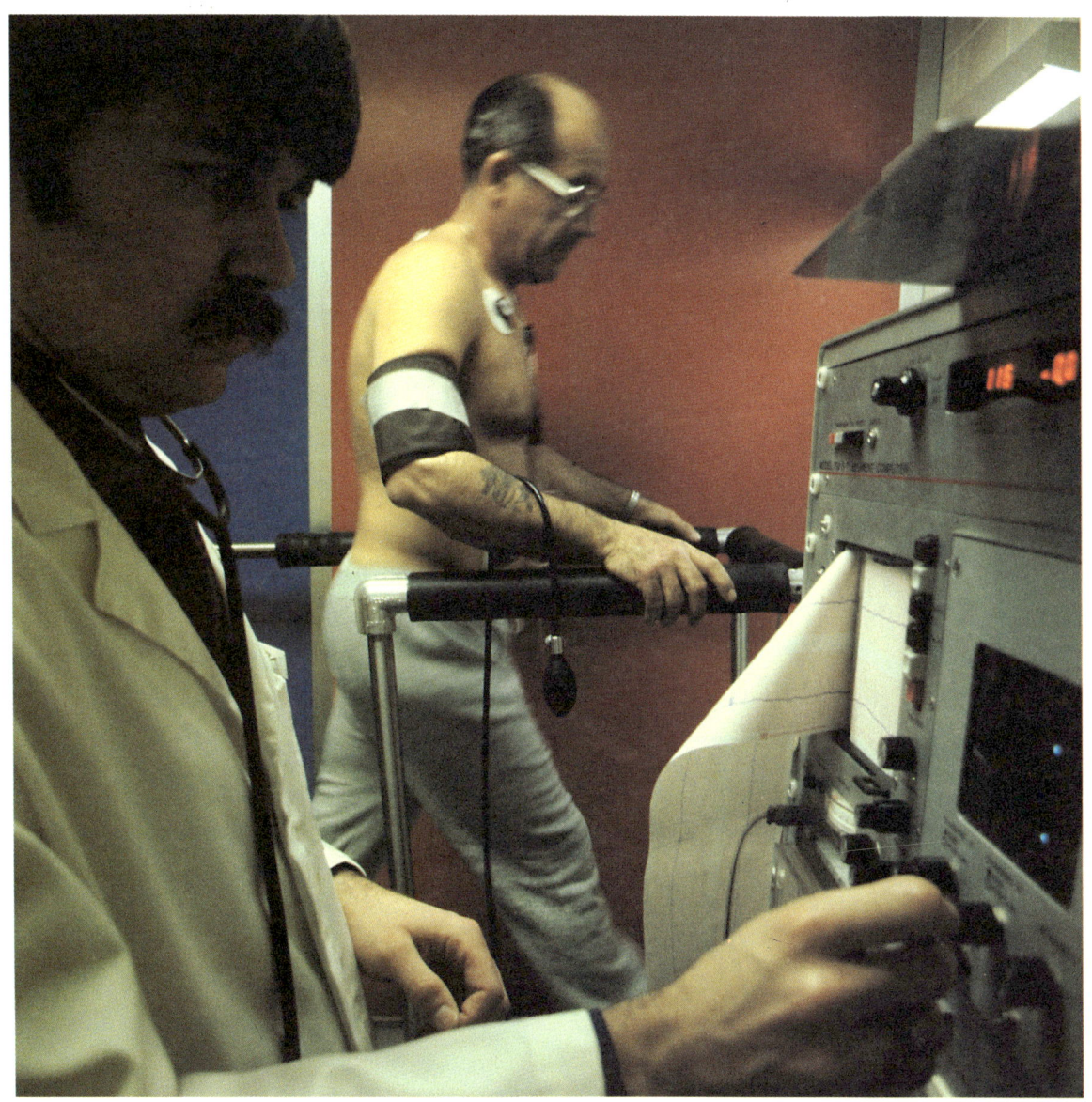
This patient's heart and lung capacity is being tested during a physical exam.

People are warned of developing air pollution through newspapers, television and radio.

12

SAMPLING AIR POLLUTION

To check outdoor air for dangerous gas and particles, local anti-pollution agencies place air **filters** made of paper or cloth fibers at different locations and heights. The filters collect samples of particles in the air as the air passes through them. Workers test these samples for pollutants. Weather bureaus make reports on radio, television and in the newspaper when heavy air pollution is developing. They warn people to stay indoors until the danger is past.

One example of a dangerous period of air pollution came after the eruption of Mount St. Helen's on May 18, 1980. Because Pacific winds carried bits of stone and harsh fumes for many miles, people had to protect themselves from breathing the polluted air. Some pockets of the ash from the eruption are still trapped in the **atmosphere** and will remain there for many years.

Natural pollen in the air from these goldenrod weeds causes pollution that can irritate the eyes and nose of people.

CHAPTER THREE

DIFFERENT TYPES OF AIR POLLUTION

In the first part of this century, lung disease was not the serious problem it is today. During the last third of the century, lung cancer has become a major killer. More people now suffer from **emphysema, asthma,** and **bronchitis.** Part of the reason that more people are getting lung diseases is that air is not as clean as it once was. A solution to this problem of dirty air is difficult. It requires the government, businesses, and individuals to look at many types of air pollution and how to decrease it.

NATURAL POLLUTION

Even before humans began causing dirty air, nature itself caused air pollution. Each year in spring and fall, plants send out **pollen** grains to reproduce. Also, at any time of the year, particles from decaying plants cause gritty, irritating dust to float through the air. These particles can scratch the delicate surface of the human eye.

To clean the eye naturally, **tear ducts** produce drops of soothing fluid. These drops dampen the eye and protect it from harm. The nose makes more mucus to trap the particles and keep them from entering the lungs. This increase in tears and mucus is sometimes a sign of allergic reactions.

For some people, allergies are a serious problem. Allergy victims suffer more from natural pollutants than most people. Their eyes may become infected. The linings of their noses and **sinuses** may expand with extra fluid. These swollen linings block air to the throat and lungs. Because breathing is difficult, allergy sufferers may have trouble sleeping. Some cannot work or play because their bodies have to work too hard to draw in enough oxygen.

15

One solution for these sufferers is to take medicines that protect the linings of the nose from swelling. Some people take regular injections to keep their bodies from over-reacting to natural pollutants such as pollen and animal fur. Others wait until the problem starts before taking pills or sprays to keep down swelling and ease the body's hard work.

DIRECT TOBACCO SMOKE

One common type of air pollution comes from the bad habits of smokers. When people burn tobacco in cigarettes, cigars, or pipes, they send out a dangerous kind of air pollution. These fumes hang in the air of homes, elevators, offices, cars, and airplanes. They cloud the air with a blue-gray haze. Windows are covered with a thin, sticky film of yellow. Because of the smokers' dependence on tobacco, everyone in the area suffers.

Many diseases can result from long years of smoking. People who light up only a few times a day as well as those who smoke heavily may have trouble breathing. Smokers may catch more colds, sore throats, and bronchitis than do nonsmokers.

Pregnant smokers may give birth to **premature** or very small babies. Premature babies are babies born too soon, before their bodies have a chance to develop completely.

Smokers may suffer permanent disabling diseases such as lung cancer, asthma, or emphysema. Some may develop **cancer** of the lip, throat, tongue, or lungs. Others may suffer heart disease or **ulcers.** If these diseases take over the lungs, the sufferers may not be able to breathe without help. They may need to carry tanks of oxygen to keep them from choking or running out of air. Even with extra oxygen, many sufferers die from these diseases.

SECONDHAND SMOKE

Scientists have learned that smokers not only harm themselves, but also others around them. This dangerous form of pollution is called **passive** or **secondhand smoke**. Because tobacco fumes are heavy, they hang in the air and cling to furniture, hair, and clothing for long periods of time. People with sensitive or weak lungs may suffer harm even though they are not smoking tobacco products. One way these sufferers can protect themselves is by staying away from secondhand smoke.

Some restaurants and waiting rooms have special places where people are allowed to smoke tobacco products. These areas may have added **ventilation** to draw the harmful gases and particles out of the air. Nonsmokers can protect themselves by moving to distant parts of the room. On long airplane trips, smokers are seated in the back. In some hospitals, factories, and stores, visitors may see signs that ask them not to smoke. These signs are posted to help nonsmokers avoid discomfort or sickness and to protect the air from becoming dirty.

A common type of air pollution comes from cigarette, cigar and pipe smokers. Many diseases result from long years of exposure to tobacco smoke.

POLLUTION FROM VEHICLES AND FACTORIES

The most common source of air pollution comes from motor vehicles. Most vehicles contain **internal combustion engines**. These engines burn fuel on the inside by mixing air with gasoline. This mixture is pushed into small spaces. There, spark plugs cause the mixture to explode. The force of the explosion causes pistons, or rods, to rise and fall. This up-and-down motion moves the wheels of cars, trucks, buses, trains, and planes.

When fuel is not completely burned in the motors of these vehicles, waste is released in the form of dirty exhaust or fumes. This heavy gas contains **carbon monoxide** and other gases. Carbon monoxide can kill people by preventing the lungs from picking up oxygen. Older cars and cars that are not kept in good running condition cause far more of these dangerous gases than new cars.

In addition to carbon monoxide, car exhaust contains carbon dioxide. This gas, when mixed with poorly burned fuel, floats high above the earth. There, it affects weather patterns. Carbon dioxide allows sunlight into the earth's atmosphere but does not allow all its heat to radiate back into space. This condition is called the **greenhouse effect**.

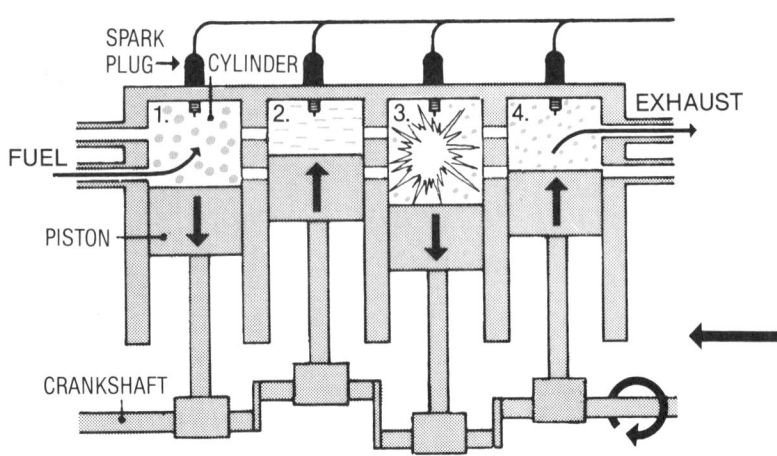

⬅ **INTERNAL COMBUSTION ENGINE**

Internal combustion engines burn fuel on the inside by mixing air with gasoline.

The greenhouse effect could turn farm lands into deserts.

Some scientists fear that the greenhouse effect will overheat the earth and cause changes in the earth's climate. Rising temperatures could cause the North Pole and South Pole ice to melt. Such a change in the earth could result in floods. The change could even turn farm land to desert. Another effect on weather could be the increase of hurricanes, drought, rain, and hail, which would damage crops and cause world hunger.

Other effects of exhaust gases include the cracking of rubber and the **corrosion** of metal. Where windows are lined with rubber insulation, polluted car exhaust can quickly eat away the seal. In factories where rubber wheels and metal parts are necessary for production, exhaust fumes can destroy an entire machine.

The second major source of pollution is large industrial factories, particularly the power plants that produce electric power. From the coal and other fuels that turn great generators comes **sulfur dioxide** and other strong gases that escape through smokestacks. These gases create serious threats to clean air.

In highly industrialized areas, the pollutants from factories are often visible by haze or discolored clouds hanging in the air. These pollutants escaping into the sky, create problems for many miles in all directions. Carried by strong winds, the pollution blankets some areas of the country, creating dirty, unpleasant breathing conditions.

The harmful effects of these air pollutants, can be greatly increased by a natural process called **inversion**. Normally dirty air moves away from the earth's surface when it is heated. Sometimes, however, a layer of warm air will stop the rise of cool air trapped under it. This inversion process holds down pollutants that might otherwise be broken up. Polluted air then remains in the sky and causes breathing discomfort for humans and animals. These conditions may last for days until changes in air pressure and moisture allow cleansing winds to clear away the pollution.

Generators from power plants can produce strong gases that escape through smokestacks, causing serious threats to clean air.

Plants, animals and forests have been affected by acid rain in the United States and many European countries.

ACID RAIN

A serious problem caused by the pollutants from car exhaust and factories is that sulfur dioxide, when mixed with rainwater, produces **acid rain**. The acid rain that falls from the sky in large cities eats away at marble buildings and statues. At first, the marble is pitted with deep holes. Eventually, whole sections of the material dissolve away. The cost of repair is very high. In the historic sections of Rome, Italy, for example, whole city blocks must be repaired before visitors can enjoy their beauty.

Acid rain also damages plants. Trees and lawns that border highways may stop growing or die from the dangerous mixture of sulfur dioxide fumes and rainwater. Vegetables or fruits from these plants may not be healthful for humans to eat. In parts of the Blue Ridge Mountains, whole forests of trees have no leaves. They've been eaten down to the bark by the deadly mixture of sulfur dioxide and rainwater.

Acid rain has killed plants and animals in lakes in the northeastern United States, eastern Canada, Norway, and Sweden. It has helped to kill forests in Germany, Austria, Poland, Czechoslovakia, Yugoslavia, Hungary, Romania, the Netherlands, and other parts of Europe and the United States.

Acid rain damages and destroys plants and trees. This tree has been tagged for removal by a red band.

INDOOR POLLUTION

Not all pollution occurs outside. Indoor pollution includes tobacco smoke, perfume, and fumes from strong detergents and cleaners. In addition to these pollutants, scientists are examining the dangers of **radon**, a harmful gas that occurs naturally in the ground and enters houses through the basement or floor. Scientists also warn of **asbestos**, a wrapping commonly used around pipes, in floors, and in ceilings of older buildings.

In addition, mold and germs grow in air conditioners and heating systems. Even fabric finishes, carpets, plastics, and building materials such as plywood contain **formaldehyde**, which irritates eyes and skin. They also contain **styrene**, a form of plastic that releases gases that can damage the liver and kidneys.

Certain buildings make people ill. This problem, called the **sick-building syndrome**, is one form of indoor pollution. Sick-building syndrome comes from poor ventilation and harmful air. A mix of chemicals, mold, germs, harmful gases, aerosols, tobacco smoke, and stale air can cause headaches, nausea, and dizziness. These symptoms may strike people in well-insulated schools, apartment houses, and offices. The problem is common in mobile homes, which are usually constructed of plastics and other materials made by people. It also occurs in certain businesses, such as dry cleaners and print shops, which use harmful chemicals every day.

In some cities, the danger of indoor pollution is many times greater than from outdoor air pollution. The problem starts when builders try to prevent heat or air conditioning loss and to protect people from bad air on the outside. By sealing and weather-stripping windows, doors, and walls, the builders also seal the indoor pollution in. These tight buildings may only make people sicker. Some workers must change jobs or move to get away from the polluted air in sick buildings and to regain their health.

Indoor air pollution can occur when there is poor ventilation in office buildings and businesses. This print shop uses harmful chemicals in their printing inks every day.

Farmers driving large tractors send out heavy exhaust fumes and create clouds of dust. This too, pollutes the air.

POLLUTION FROM FARMS

It may surprise you to learn that farms, too, are major polluters. When farmers drive heavy tractors over the earth, their engines send out heavy exhaust fumes. Their wheels and plows churn up clouds of dust. As the farmers' tools remove seeds from stalks, they loosen particles into the air.

Even worse, when farmers control weeds and insects with chemical spray, they release deadly liquids into the air. These dangerous sprays are known as **herbicides** and **insecticides**. This type of air pollution can travel on the wind to woodlands, waterways, and homes.

When these insecticides or herbicides reach trees or other plants, they can weaken or kill the plants. These chemicals also kill bees, which are necessary for the growth of flowers and vegetables. If chemical sprays reach nesting grounds, they can destroy birds, which are one of nature's most effective controllers of insects.

Harmful herbicides and insecticides can travel even farther on land. These chemicals enter the soil as hazardous **runoff** and can pollute streams and creek banks for miles. The chemicals can also poison **ground water**, which is a major source of the water we drink.

Bees which are necessary for the growth of flowers and vegetables are sometimes killed by insecticides.

OTHER POLLUTERS

Many other businesses also cause harm to the air. For example, **strip mining** is the process of taking minerals from the surface of the earth. Strip mining creates air pollution by making large clouds of dust and dirt. Some of these particles are ordinary dirt. Others are dangerous substances, such as **heavy metals**, which can cause lung cancer.

To limit the air pollution caused by strip mining, mine owners can use a watering system to dampen the dust and heavy metals, causing them to fall back to earth. Mine owners can also protect their neighbors by enclosing the belts that carry metals to trucks. This keeps the dust from being thrown into the air. Instead, the dust particles then collect in heaps alongside the mine.

A power shovel strips away soil and rock to expose a coal seam just below the surface.

Strip mining creates clouds of dust which often contain dangerous substances such as heavy metals.

Composting leaves, branches, and grass rather than burning them helps to control air pollution. The resulting rich compost can improve soil and stop erosion.

CHAPTER FOUR

CLEANING THE AIR

Once pollutants reach the air, it is almost impossible to remove them. Obviously, it is much better for all living things on earth if humans prevent inversion and other types of air pollution before they start. The best way to prevent dirty air is to control the causes of the problem. Education can help citizens learn how to make the **environment** cleaner.

STOPPING AIR POLLUTION AT HOME

Citizens often point their fingers at factories as the only source of air pollution. But individuals, too, have a responsibility to stop harmful particles from dirtying the air. By reducing waste, taking trash to a landfill, recycling usable materials, and **composting** leaves and branches rather than burning them, citizens help stop a major cause of smoky air.

Also, the resulting rich compost can improve soil and stop erosion, another environmental problem. Gardeners can limit **pesticide** spraying or spray only on windless days to stop the drift of poisons to their neighbors.

One source of air pollution from homes is the smoke from fireplace chimneys. People should make certain that chimneys and flues are free from soot buildup. An annual cleaning gets rid of soot and squirrel nests. This cleaning improves the proper intake of air and allows a free flow of smoke from the chimney. Also, burning only clean, dry logs and kindling is safer than using the fireplace or woodstove as a garbage incinerator. Fire builders should never burn treated lumber, colored or metallic paper, plastic foam, dangerous chemicals, or plastics.

MAKING OUR NATION A BETTER NEIGHBOR

Uncontrolled air pollution is a serious threat to relationships with other countries. For example, because the industrial smoke of northern American cities escapes into the air, it also threatens the health of people in Canada. Canadians are naturally upset that American industry is polluting their air. Acid rain is destroying their forests and killing their fish. Such a loss endangers the whole economy of Canada, which depends heavily on the wood industry. In addition, acid rain discourages visitors who come to fish and enjoy the countryside.

A similar situation affects our Mexican neighbors. Harmful gases from **refineries** in Texas and Louisiana affect the farms of Mexico. The fruits and vegetables poisoned by pollution endanger the Mexicans. These products are also purchased in the United States. Thus, the pollutants return to our own citizens and endanger the Americans who buy and eat them.

Acid rain can damage the forests and fishing industries, thus, endangering the economy of a nation.

Harmful gases released into the air from oil refineries can poison the fruits and vegetables that we eat.

REDUCING POLLUTION FROM CARS

Since the late 1960s, government agencies have enacted laws that help control exhaust. For instance, new cars, buses, and trucks cannot be sold without **catalytic converters,** which are devices that change harmful gases into safer exhaust. These converters raise the price of new vehicles as well as the cost of operation, but at the same time, they help control air pollution.

At one time, motor vehicles used gasoline that had lead in it. Lead is a dangerous heavy metal. When lead enters the lungs while breathing, it can poison the body. Lead poisoning can cause severe problems to the brain. Because lead poisoning is so deadly, laws now require people to buy gasoline that is lead-free.

Highway planners are also helping reduce air pollution by designing highways to keep traffic moving. When cars move smoothly and evenly through traffic, they spend less time standing still with the motor running. Therefore, the cars release less pollution into the air.

Changing the type of fuel we use would be a great help to stopping pollution from vehicles. If drivers could find a clean, low-priced fuel, they could help solve the problem of air pollution. One possible fuel is a form of alcohol made from corn. In addition to solving an environmental problem, this change would give farmers a new market for the corn they produce.

MAKING SAFER FACTORIES

Factories and power plants, like motor vehicles, must burn fuels to operate. Because large factories use great amounts of coal, gas, and other raw materials each day, they produce tons of polluted gases. One of the most unpleasant of these gases results from the **sulfur** that is found in coal. When sulfur forms hydrogen sulfide, it gives off a smell like rotten eggs. When it forms sulfur dioxide, it becomes a poisonous gas.

Highway planners are helping to reduce air pollution by designing highways to keep traffic flowing smoothly. The use of signal lights at freeway entrances help stop-and-go congestion.

35

When factory and power plant owners use less harmful solids, liquids, and gases, fewer pollutants enter the air.

Some of these gases can be made harmless by the addition of water or chemicals. Others, however, are directed into the air through factory smokestacks. According to clean air laws, smokestacks must be taller than the surrounding buildings or hills. From this height, the smoke rises above the nearby residents and workers.

In addition, to halt heavy particles or **fly ash,** smokestacks can be covered with caps that trap solids and keep them from mixing with the air. Then factory owners dispose of the ash in **sanitary landfills.**

Many other up-to-date devices can help cut down the number of particles escaping into the air. Some smokestacks are equipped with a **flare tip** to reduce the amount of pollution that escapes into the air. This device injects steam into the gas as it leaves the smokestack. The addition of steam helps reduce the amount of smoke. Sometimes the steam even eliminates pollutants entirely.

Another air pollution control device magnetizes the particles. Like magnets, the particles stick together and fall harmlessly out of the way of escaping smoke. Later, factory workers gather the particles and dispose of them safely in sanitary landfills.

Many factories have **recovery systems** that actually vacuum the gas as it escapes. These systems cool the hot gas, remove valuable chemicals, and collect them in barrels. Then these chemicals can be reused. Such systems cut down not only pollution, but also expenses.

Some smokestacks contain **scrubbers,** which are devices that remove the harmful particles from factory exhaust. These devices help reduce the pollution from factories even more. By lining the upper section of the **flue** with **activated charcoal,** factory owners can trap harmful particles and change them into safer **by-products.**

Harmful factory exhaust can also be reduced by controlling raw materials. When factory owners substitute less harmful solids, liquids, and gases, they cause fewer pollutants to enter the air. In this way, factory owners become good neighbors to the people who live close to their buildings.

Help do your part to save the environment by walking short distances rather than going by car.

CHAPTER FIVE

THE INDIVIDUAL'S PART

Their trip to California made Jen and Marty Graham aware of air pollution. They became so concerned that their mother took them to visit a local environmental specialist.

The specialist told the Grahams that the average person cannot stop air pollution singlehandedly. However, each person can be more responsible for activities that cause dangerous particles and fumes to cloud clean air and harm plants and animals. Here is a list of suggestions that the environmental specialist gave the Grahams:

PROTECT THE AIR

1. Don't buy products in **aerosol** spray cans. Instead, select products that come in pump sprayers.
2. For short distances, travel on foot, by bus, or by bicycle. If you must go by car, combine several errands on one trip.
3. Start a **carpool** in your area. Share rides to school and community activities. Shop with a friend or neighbor.
4. Burn only dry hardwoods in fireplaces. Avoid using the fireplace or woodstove as a garbage incinerator. Keep chimneys clean.
5. Keep vehicles in good condition by having frequent engine tune-ups. A clean engine burns fuel more efficiently.
6. Consider alternate fuels for vehicles, such as batteries, propane, methane, ethanol, or alcohol.

PROTECT YOURSELF

1. Never pick berries, mushrooms, or other edible plants along a highway or near a factory where pollution or insecticide and herbicide sprays may have coated them.
2. Use paint, spot removers, and other products that release harmful fumes in open air. If you must use them indoors, turn on an exhaust fan and open windows.
3. Stay away from smokers. If you must ride in a car or stay in a room where people smoke, keep windows open.
4. If you are sensitive to polluted air, use a mask for lawn mowing, painting, and other chores that pollute the air you breathe.
5. Listen to radio broadcasts for news of heavy air inversion or smog in your area. When outside breathing conditions are poor stay indoors.
6. Use charcoal grills outdoors and far enough away from houses to prevent strong fumes from bothering other people. Never burn charcoal in an enclosed area.
7. Study science to help you understand the complex materials that pollute the air.
8. Grow plants that produce oxygen in your home.
9. Buy products that are free of fragrance to prevent extra pollutants from harming your lungs and skin.

Use charcoal grills away from other people to prevent strong fumes from irritating their eyes and nose.

ENCOURAGE GOVERNMENT OFFICIALS TO TAKE ACTION

1. Write or call local officials or state or national representatives. Encourage them to vote for stronger anti-pollution laws.
2. Support citizens groups, such as those that police factories, power plants, airports, and strip mining operations. Insist on heavy fines for polluters.
3. Observe the air in your area. If you see large clouds of pollutants rising from factories, call a representative of the **Environmental Protection Agency (EPA)** near you.
4. Ask questions. Read articles about pollution. By learning more, you can be an effective voice against air pollution.

ENCOURAGE INDUSTRY TO KEEP THE AIR CLEAN

1. Avoid products made by manufacturers who pollute or who refuse to cooperate in the cleanup and protection of air.
2. Write thank-you notes to companies that are making an effort to keep the air clean.

Write to your local officials to support anti-pollution laws and to report any unusual pollutants you see.

GLOSSARY

acid rain (A sihd RAYN)   rain that mixes with sulfur dioxide and nitrogen oxide and is more acidic than normal rain

activated charcoal (AK ti vay tind CHAHR kohl)   charcoal that has many pores, or openings, to trap harmful particles

aerosol (AYR uh sahl)   liquid sprayed from a can that contains a harmful pressurized gas

allergies (A luhr jeez)   extreme physical reactions, such as coughing, sneezing, rashes, or nausea

asbestos (as BEHS tohs)   a cancer-causing fiber that is used in insulation

asthma (AZ muh)   a narrowing of airways that causes breathing difficulty, wheezing, and coughing

atmosphere (AT muh sfeer)   the layer of gases that surrounds the earth

bronchitis (brahn KY tihs)   a disease that causes watery swellings of the tubes that lead to the lungs

by-products (BY prahd uhkts)   waste products made by a factory while it is creating useful goods

cancer (KAN suhr)   an abnormal growth in cells

carbon dioxide (KAR buhn dy AHK syd)   a gas that the body produces when it burns oxygen

carbon monoxide (KAR buhn muhn AHK syd)   a deadly, odorless gas that results from car exhaust and the incomplete burning of fossil fuels

carpool (KAR pool)   a group of people sharing a ride in the same vehicle

catalytic converters (kat uh LIHT ihk kuhn VUHRT uhrz)   devices in vehicles that clean the exhaust fumes

composting (KAHM pohs tihng)   collecting dead leaves and other plant material and encouraging them to decay

corrosion (kuh ROH zyuhn)   the chemical destruction of metal

emphysema (ihm fuh ZEE muh)   a disease that causes the sacs in the lungs to break down

environment (ihn VYRN mihnt)   anything that influences life and makes living possible

Environmental Protection Agency (EPA)   a federal agency that helps protect the environment of United States citizens

filters (FIHL tuhrz)   fine screens that separate heavy particles from liquids or gases

flare tip (FLAYR TIHP)   a device in a smokestack that helps stop particles from escaping by mixing them with steam

fly ash (FLY ASH)   solid particles of dust, soot, and ash which are carried away from a fire

flue (FLOO)   a tube in a chimney, through which smoke is drawn off

formaldehyde (fohr MAL dih hyd)   a colorless gas used in cleaners and as a preserver

greenhouse effect (GREEN hows eh FAHKT)   the trapping of the sun's heat within the earth's atmosphere

ground water (GROWND wah tuhr)   water that collects below the earth's surface

heavy metals (HEHV ee MEH tuhlz)   metals, including zinc, mercury, lead, chromium, and arsenic, which can poison humans and animals

herbicide (UHR bih syd)   a chemical that kills plants

insecticide (ihn SEHK tih syd)   a chemical that kills insects

internal combustion engine (IHN tuhr nuhl kuhm BUHS chyuhn IHN jihn)   a type of engine that burns fuel

inversion (ihn VUHR zyuhn)   a blanket of pollution mixed with wet air that remains trapped near the earth and makes breathing difficult

mucous linings (MYOO kuhs LY nihngz)   body linings made of mucus

mucus (MYOO kuhs)   a sticky substance made by the body; helps remove dirt and smoke from the air a person breathes

oxygen (AHK sih jihn)   a gas that is necessary to human and animal life

passive smoke (PA sihv SMOHK)   the breathing of tobacco fumes that a smoker has exhaled

pesticides (PEHST ih sydz)   chemicals that kill unwanted insects, plants, and fungus

pollen (PAHL ihn)   a fine dust produced by plants

pollutants (puh LOO tuhnts)   airborne particles that cause pollution

pollution (puh LOO shuhn)   dirtying of natural resources

premature (pree muh TYOOR) born too soon

radon (RAY dahn) a harmful gas that occurs naturally in soil and rocks

recovery systems (ree KUHV ree SIHS tuhmz) devices that take chemicals out of the air and return them to storage for later use

refineries (ree FYN reez) factories that change crude oil into gasoline and other products

runoff (RUHN ahf) rapid movement of water that picks up soil and carries it away

sanitary landfill (SAN in tah ree LAND fihl) a place where solid waste is buried between layers of dirt

scrubbers (SKRUHB uhrz) devices in smokestacks that collect harmful particles and keep them from dirtying the air

secondhand smoke see passive smoke

sick-building syndrome (SIHK BIHL dihng SIHN drohm) sickness caused by harmful air in a tightly-constructed building

sinuses (SY nuh sez) chambers in the skull that contain air

smog (SMAHG)   a mixture of smoke and fog that hovers over the earth when the weather is hot and wet

strip mining (STRIHP MY nihng)   a mining operation that strips away the soil that covers a valuable mineral

styrene (STY reen)   a chemical used to make plastic and rubber

sulfur (SUHL fuhr)   a chemical that occurs naturally in coal and other minerals

sulfur dioxide (SUHL fuhr dy AHK syd)   a by-product of car exhaust that forms acid rain

tear ducts (TEER DUHKTS)   the parts of the eye that produce a protective liquid

ulcers (UHL sehrz)   open sores in the stomach lining

ventilation (vihn tuhl AY shuhn)   providing enough air to breathe